HYDROELECTRICITY

EDITED BY ELIZABETH LACHNER

Published in 2019 by Britannica Educational Publishing (a trademark of Encyclopædia Britannica, Inc.) in association with The Rosen Publishing Group, Inc.
29 East 21st Street, New York, NY 10010

Distributed exclusively by Rosen Publishing.
To see additional Britannica Educational Publishing titles, go to rosenpublishing.com.

First Edition

Britannica Educational Publishing
J.E. Luebering: Executive Director, Core Editorial
Andrea R. Field: Managing Editor, Compton's by Britannica

Rosen Publishing
Amelie von Zumbusch: Editor
Brian Garvey: Series Designer/Book Layout
Cindy Reiman: Photography Manager
Sherri Jackson: Photo Researcher

Library of Congress Cataloging-in-Publication Data

Names: Lachner, Elizabeth, editor.
Title: Hydroelectricity / edited by Elizabeth Lachner.
Description: New York : Britannica Educational Publishing, in Association with Rosen Educational Services, 2019. | Series: Exploring energy technology | Includes index. | Audience: Grades 5–8.
Identifiers: LCCN 2018009225| ISBN 9781508106197 (library bound) | ISBN 9781508106180 (pbk.)
Subjects: LCSH: Hydroelectric power plants—Juvenile literature. | Water-power—Juvenile literature.
Classification: LCC TK1081 .H8935 2019 | DDC 621.31/2134—dc23
LC record available at https://lccn.loc.gov/2018009225

Manufactured in the United States of America

Photo credits: Cover MarianneBlais/E+/Getty Images; cover, back cover, pp. 1, 3, 4-5 (background), 9, 19, 28, 35, 42, 44, 47 © iStockphoto.com/tolokonov; pp. 4-5 Harald Sund/Photographer's Choice/Getty Images; p. 7 Camerique/Archive Photos/Getty Images; p. 10 Monty Rakusen/Cultura/Getty Images; pp. 11, 15, 39 © Encyclopædia Britannica, Inc.; pp. 12, 26, 30, 31, 40 (background) sakkmesterke/Shutterstock.com; p. 13 SF Photo/Shutterstock.com; p. 16 DEA/C. Sappa/De Agostini/Getty Images; p. 18 © P123; p. 20 ManeeshUpadhyay/Shutterstock.com; p. 22 Marekuliasz/Shutterstock.com; p. 23 Bloomberg/Getty Images; p. 25 Pulsar Imagens/Alamy Stock Photo; p. 27 Romaset/Shutterstock.com; p. 29 © Courtesy of the Bibliothèque Nationale, Paris; p. 31 © Scott Latham/Fotolia; p. 33 Piotr Wytrazek/iStock/Thinkstock; p. 34 Westend61/Getty Images; p. 37 Frans Lanting/Mint Images/Getty Images; p. 38 © ftfoxfoto/Fotolia; p. 41 Mauro Repossini/iStock/Thinkstock.

CONTENTS

INTRODUCTION

The roar of a waterfall suggests the power of water. Rampaging floodwaters can uproot strong trees and twist railroad tracks. When the power of water is harnessed, however, it can do useful work for humans.

Since ancient times people have put the energy in the flow of water to work. They first made water work for them with the waterwheel—a wheel with paddles around its rim. Flowing water rotated the waterwheel, which in turn ran machinery that was linked to it. Today, new kinds of waterwheels—known as turbines—spin generators that produce electricity. Electricity from water-turned generators is known as hydroelectricity.

By building a dam across a river, the natural upstream water level is elevated and a difference

These turbines produce electricity at the Grand Coulee Powerplant, in Grand Coulee, Washington.

in head is created that can be used to drive turbines and generate electricity. A large upstream reservoir may balance seasonal water flow; rain or melted snow can be stored in the reservoir during the wet season to provide electricity during dry seasons. Small "run-of-river" reservoirs or ponds are not large enough to provide seasonal balance. They can, however, provide extra power during daily peak hours.

In most communities, electric-power demand varies considerably at different times of the day. To even the load on the generators, pumped-storage hydroelectric stations are occasionally built. During off-peak periods, some of the extra power available is supplied to the generator operating as a motor, driving the turbine to pump water into an elevated reservoir. Then, during periods of peak demand, the water is allowed to flow down again through the turbine to generate electrical energy. Pumped-storage systems are efficient and provide an economical way to meet peak loads.

Waterpower is distributed unevenly among the continents and nations of the world. Europe and North America have developed much of their waterpower. Asia, South America, and Africa have abundant waterpower potential, but while countries such as China and Brazil have become leading hydroelec-

The Grand Coulee Dam was built between 1933 and 1941. Its water is used to generate electricity at the Grand Coulee Powerplant.

tric producers, much of the waterpower resources on those continents remains undeveloped.

The use of hydropower in the United States has been significant since about 1900. The Columbia River drainage area, which is the site of the Grand Coulee, Bonneville, and Hungry Horse dams, has both the greatest potential and the greatest developed waterpower. California ranks second in the nation in potential waterpower, while the Ohio River basin is second in developed waterpower. Significant additional waterpower potential still exists in the Missouri and Ohio River basins and in the North and South Atlantic portions of the United States. The largest hydroelectric power plant in the United States is the Grand Coulee, which has a capacity of 7,600 megawatts.

Among water's virtues as a source of power is the fact that it is renewable, meaning that it can be used again after it supplies energy. This contrasts with energy sources—including fossil fuels such as coal, petroleum, and natural gas—that are destroyed when they are used. Hydroelectric installations also supply added benefits to a region. For example, a dam built to provide a head for water turbines usually creates a reservoir that can supply water for irrigation and drinking. Potential sources of waterpower include the tides, waves, and currents of the ocean, as well as the differences in temperature between the surface and the bottom of the ocean.

CHAPTER ONE

TYPES OF WATERPOWER

The vast majority of the world's hydroelectricity is produced by the power of falling water. Hydroelectric power plants are usually located in dams that are built across rivers. In a dam water is collected at a higher elevation and is then led downward through large pipes to a lower elevation. The falling water causes the water turbines to rotate, producing electricity.

Oceans can also be used to create hydroelectricity. Those waterpower sources are known as tidal power and wave power. Tidal power is created during the tide, when the water level along the oceanic coast changes. Wave power is harnessed by the up-and-down motion of waves.

HYDROELECTRIC PLANTS

In the generation of hydroelectric power, water is collected or stored at a higher elevation and led downward through large pipes or tunnels (called penstocks) to a lower elevation. The difference in these two elevations is known as the head. At the end of its passage down the pipes, the falling water causes turbines to rotate. The turbines in turn drive generators, which convert the turbines' mechanical energy into electricity. Transformers are then used to convert the alternating voltage suitable for the generators to a higher voltage suitable for long-distance

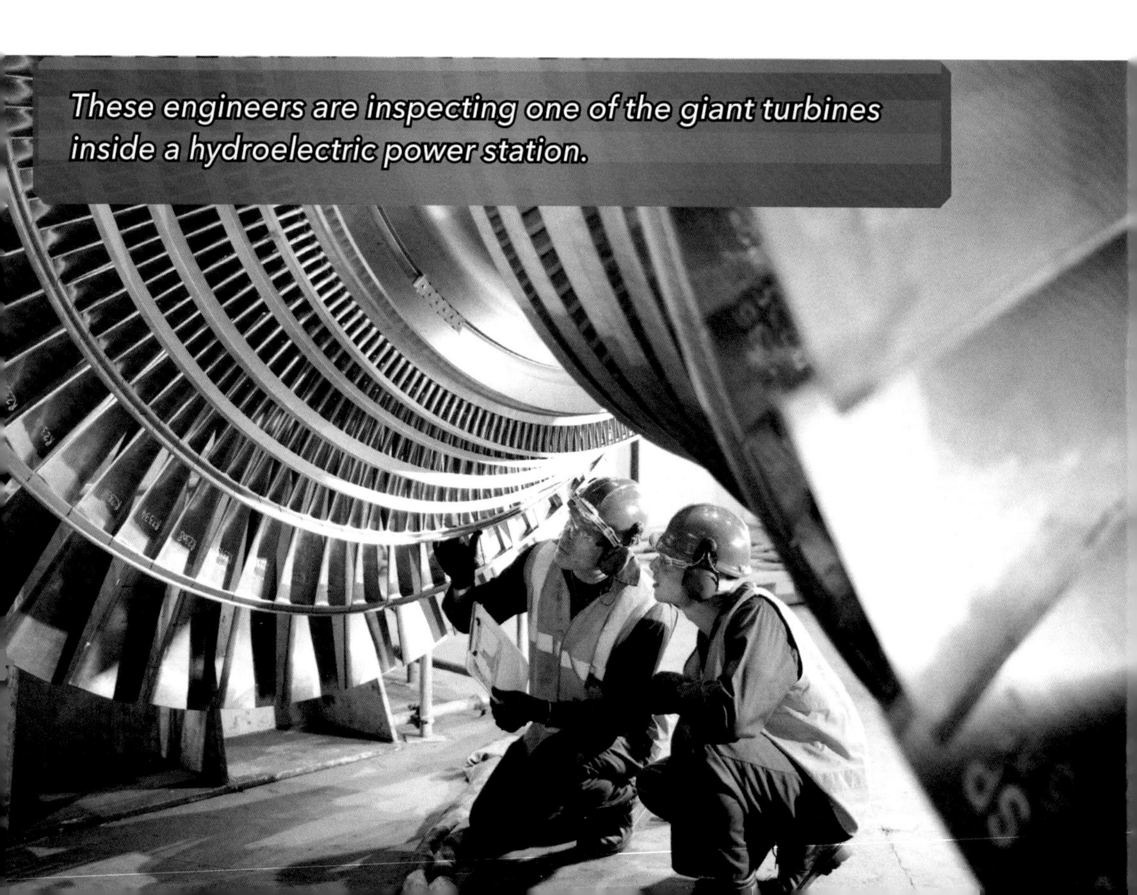

These engineers are inspecting one of the giant turbines inside a hydroelectric power station.

transmission. The structure that houses the turbines and generators, and into which the pipes or penstocks feed, is called the powerhouse.

Hydroelectric power plants are usually located in dams that impound rivers, thereby raising the level of the water behind the dam and creating as high a head as is feasible. The potential power that can be derived from a volume of water is directly proportional to the working head, so that a high-head installation requires a smaller volume of water than a low-head installation to produce an equal amount of power. In some dams, the powerhouse is constructed on one flank of the dam, part of the dam being used

This diagram of the Hoover Dam shows the dam itself and part of the reservoir behind it. The reservoir, known as Lake Mead, is one of the largest man-made lakes in the world.

HOW TURBINES WORK

The hose that firefighters drag to a burning building is filled with water almost to the bursting point. The nozzle, however, is turned off. One person can hold it easily. When the nozzle is opened, the big stream starts to spurt. The hose straightens and jumps like a giant snake. Firefighters struggle to control it. That kind of power, coming from the flow of a fluid, is used in turbine engines.

If someone wanted to use the tremendous power within the firefighter's hose, it could be done in two ways. A wheel could be set up with vanes on its rim, and the full force of the water could be directed against these vanes, making the wheel spin at great speed. This would be an example of an impulse turbine, the simplest form of turbine.

On the other hand, imagine a wheel shaped like a windmill wheel inside the hose. It would cause the water to slow while the water made it turn. It would not turn as fast as the one described above, which receives the full force of the stream all at once, but it would deliver a steadier and equally powerful motion. This would be an example of the reaction turbine.

as a spillway over which excess water is discharged in times of flood. Where the river flows in a narrow steep gorge, the powerhouse may be located within the dam itself.

Water turbines are of immense value for generating electricity where a swift and plentiful supply of water is available. Some of the largest hydroelectric plants in the world are situated at Niagara Falls (on the United States–Canadian border), where the vast flow is harnessed by mammoth reaction turbines of the Francis type. In the high Sierra Nevada in California, where streams can be dammed to get high pressures, impact turbines called Pelton wheels are

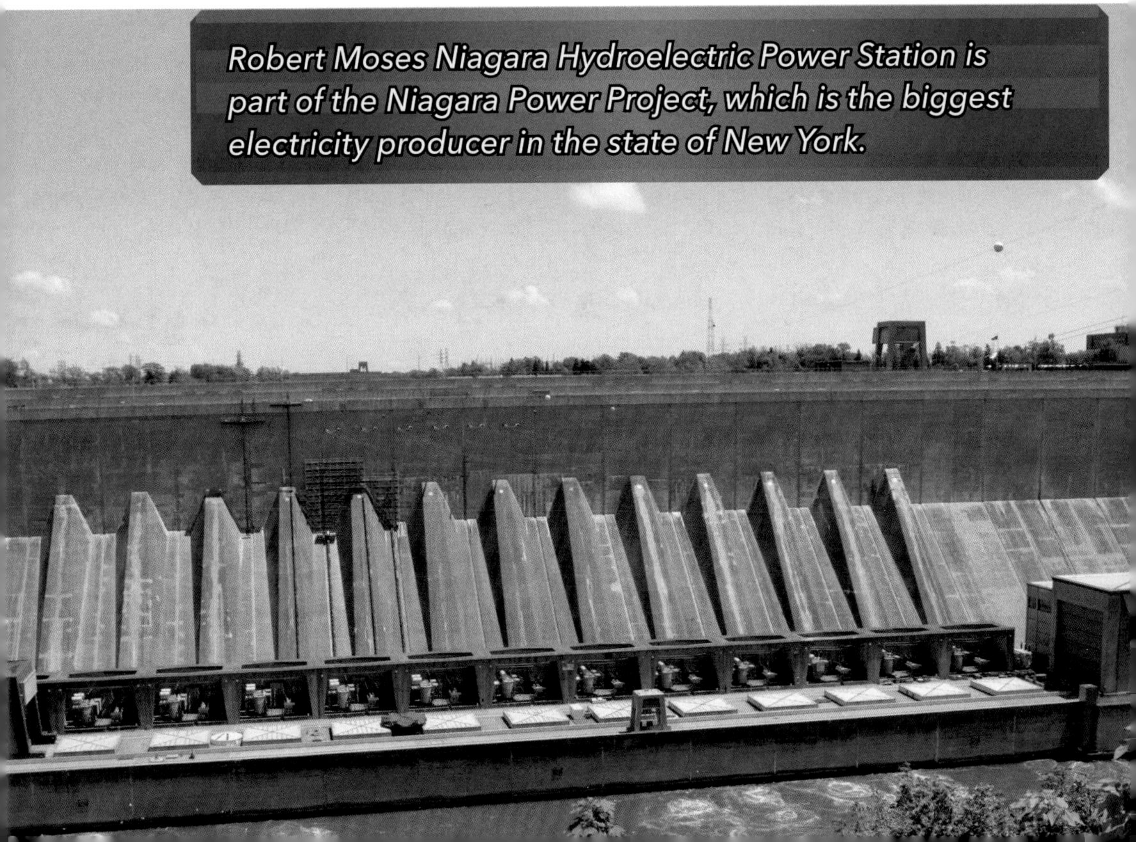

Robert Moses Niagara Hydroelectric Power Station is part of the Niagara Power Project, which is the biggest electricity producer in the state of New York.

favored. Where the volume of water is great but pressure is low, as at Louisville, Kentucky, on the Ohio River, reaction turbines of the propeller type are used.

PUMPED HYDRO STORAGE

The mechanism of pumped storage can be compared to a giant storage battery in which mechanical rather than electric power is stored intermittently. During the off-peak hours excess electricity produced by nuclear or coal- or oil-fired plants is used to pump water from a lower- to a higher-level reservoir for storage. During peak hours the water is allowed to flow down between the reservoirs through turbines in order to generate electricity. The hydraulic units are designed to operate as pumps when rotating in one direction and as turbines when rotating in the opposite direction. Similarly, motors that drive the pumps can be reversed to act as generators.

TIDAL POWER

At any location the surface of the ocean oscillates between high and low points, called tides, about every 12½ hours. In certain large bays, this tidal action can be greatly amplified. It can also create waves that move at speeds of up to 60 feet (18 meters) per second.

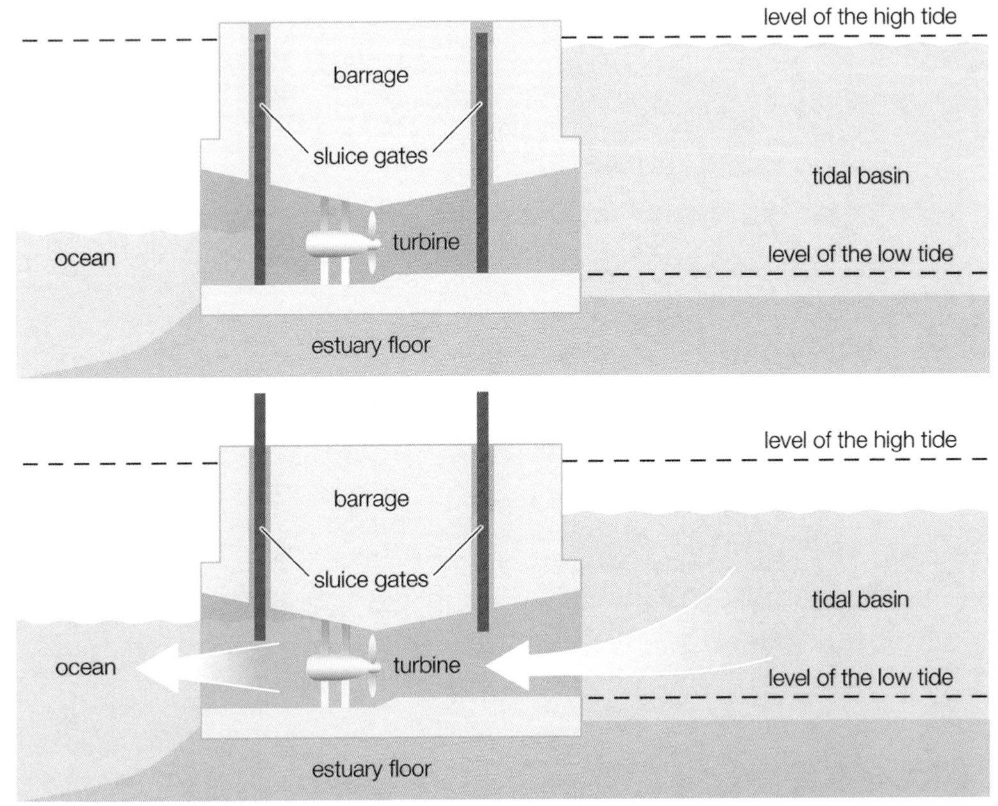

This diagram explains how a tidal barrage works. Tides actually occur in all bodies of water but are seen most prominently where the oceans meet the land as well as in bays and harbors.

In order to tap the potential energy of tides, a barrage with a gate is built across the mouth of an estuary. As the tide rises, water flows in through the open gate; when the tide begins to recede, the gate is closed. The trapped water is then channeled through a turbine. Normally such plants can generate electricity for about five hours, with seven hours of standstill and refilling. A two-way system that extracts energy from the tide as it flows both in and out of the barrage may also be used.

Thus far tidal power plants have not been economically feasible as a primary electricity source because of their intermittent power generation and their high construction costs. Tidal power may be more useful, however, for intermittent peak power generation. In this case, as in pumped-storage systems, the same machines alternate as pumps and turbines, but the head gained from the tide is used

The La Rance Tidal Power Station took about six years to build and began operation in 1966. It uses twenty-four turbines.

to reduce pumping power during high tide and to increase turbine output during low tide. The Rance River estuary in Brittany, France, is one area where hydroelectric power plants have been constructed to take advantage of the rise and fall of tides.

WAVE ENERGY

The forward motion of a wave represents its kinetic energy, and many devices have been proposed for the capture of this energy. The vertical distance between the crest and valley of a wave represents the wave's potential energy. A simple device for exploiting this potential energy consists of an air-filled floating tank, open at the bottom and attached at the top to an air-driven turbine. As the air in the tank is compressed by a wave, the air is forced through the turbine, producing power. As the wave falls, the pressure within the tank drops, the turbine valve closes, and air is then readmitted into the tank through another valve.

The areas of greatest potential for wave energy development are in the latitudes with the highest winds (latitudes 40°–60° N and S) on the eastern shores of the world's oceans. For instance, the world's first operational wave-power generator is located off the coast of Aguçadora, Portugal, producing as

This is one of the three machines used to generate power from waves off the coast of Aguçadora, Portugal.

much as 2.25 megawatts from three huge jointed tubes that float on the surface of the Atlantic Ocean. Individual power generators are located at the tubes' joints and activated by wave motion. In addition, a large potential for wave power systems exists in the British Isles and the Pacific Northwest of the United States. According to some estimates, wave energy has the potential to supply approximately 10 percent of global electricity production.

CHAPTER TWO

WATER TURBINES

Water turbines are generally divided into two categories: (1) impulse turbines used for high heads of water and low flow rates and (2) reaction turbines normally employed for heads below about 1,475 feet (450 m) and moderate or high flow rates. Turbines can be arranged with either horizontal or, more commonly, vertical shafts. Wide design variations are possible within each type to meet the specific local hydraulic conditions. Today, most hydraulic turbines are used for generating electricity in hydroelectric installations.

IMPULSE TURBINES

In an impulse turbine the potential energy, or the head of water, is first converted into kinetic energy by

For a large height difference, or head, and low flow rates, the Pelton wheel or turbine is used. The water flowing down a pipe, or penstock, accelerates through a nozzle at the bottom.

discharging water through a carefully shaped nozzle. The jet, discharged into air, is directed onto curved buckets fixed on the periphery of the runner to extract the water energy and convert it to useful work.

The power of a given wheel can be increased by using more than one jet. Two-jet arrangements are common for horizontal shafts. Sometimes two separate runners are mounted on one shaft driving a single electric generator. Vertical-shaft units may have four or more separate jets.

If the electric load on the turbine changes, its power output must be rapidly adjusted to match the demand. This requires a change in the water flow rate to keep the generator speed constant. The flow rate through each nozzle is controlled by a centrally located, carefully shaped spear or needle that slides forward or backward as controlled by a hydraulic servomotor.

Another type of impulse turbine is the turgo type. The jet impinges at an oblique angle on the runner from one side and continues in a single path, discharging at the other side of the runner. This type of turbine has been used in medium-sized units for moderately high heads.

REACTION TURBINES

In a reaction turbine, forces driving the rotor are achieved by the reaction of an accelerating water flow in the runner while the pressure drops. The reaction principle can be observed in a rotary lawn sprinkler where the emerging jet drives the rotor in the opposite direction. Because of the great variety of possible runner designs, reaction turbines can be used over a much larger range of heads and flow rates than impulse turbines. Reaction turbines typically have a

For low heads and high flow rates, vertically installed propeller or Kaplan turbines are used. Unlike with ships, where the propeller moves the water, propeller turbines use water to turn the propeller.

spiral inlet casing that includes control gates to regulate the water flow. In the inlet a fraction of the potential energy of the water may be converted to kinetic energy as the flow accelerates. The water energy is subsequently extracted in the rotor.

There are four major kinds of reaction turbines in wide use: the Kaplan, Francis, Deriaz, and propeller type. In fixed-blade propeller and adjustable-blade Kaplan turbines (named after the Austrian inventor

Victor Kaplan), there is essentially an axial flow through the machine. The Francis- and Deriaz-type turbines (after the British-born American inventor James B. Francis and the Swiss engineer Paul Deriaz, respectively) use a "mixed flow," where the water enters radially inward and discharges axially. Runner blades on Francis and propeller turbines consist of fixed blading, while in Kaplan and Deriaz turbines the blades can be rotated about their axis, which is at right angles to the main shaft.

Hydroelectric plants often use Francis turbines with adjustable blades. The angle of such blades can be changed to increase a turbine's efficiency.

OUTPUT AND SPEED CONTROL

If the load on the generator is decreased, a turbine will tend to speed up unless the flow rate can be reduced accordingly. Similarly, an increase of load will cause the turbine to slow down unless more water can be admitted. Since electric-generator speeds must be kept constant to a high degree of precision, this leads to complex controls. These must take into account the large masses and inertias of the metal and the flowing water, including the water in the inflow pipes (or penstocks), that will be affected by any change in the wicket gate setting. If the inlet pipeline is long, the closing time of the wicket gate must be slow enough to keep the pressure increase caused by a reduction in flow velocity within acceptable limits. If the closing or opening rate is too slow, control instabilities may result. To assist regulation with long pipelines, a surge chamber is often connected to the pipeline as close to the turbine as possible. This enables part of the water in the line to pass into the surge chamber when the wicket gates are rapidly closed or opened. Medium-sized reaction turbines may also be provided with pressure-relief valves through which some water can be bypassed automatically as the governor starts to close the

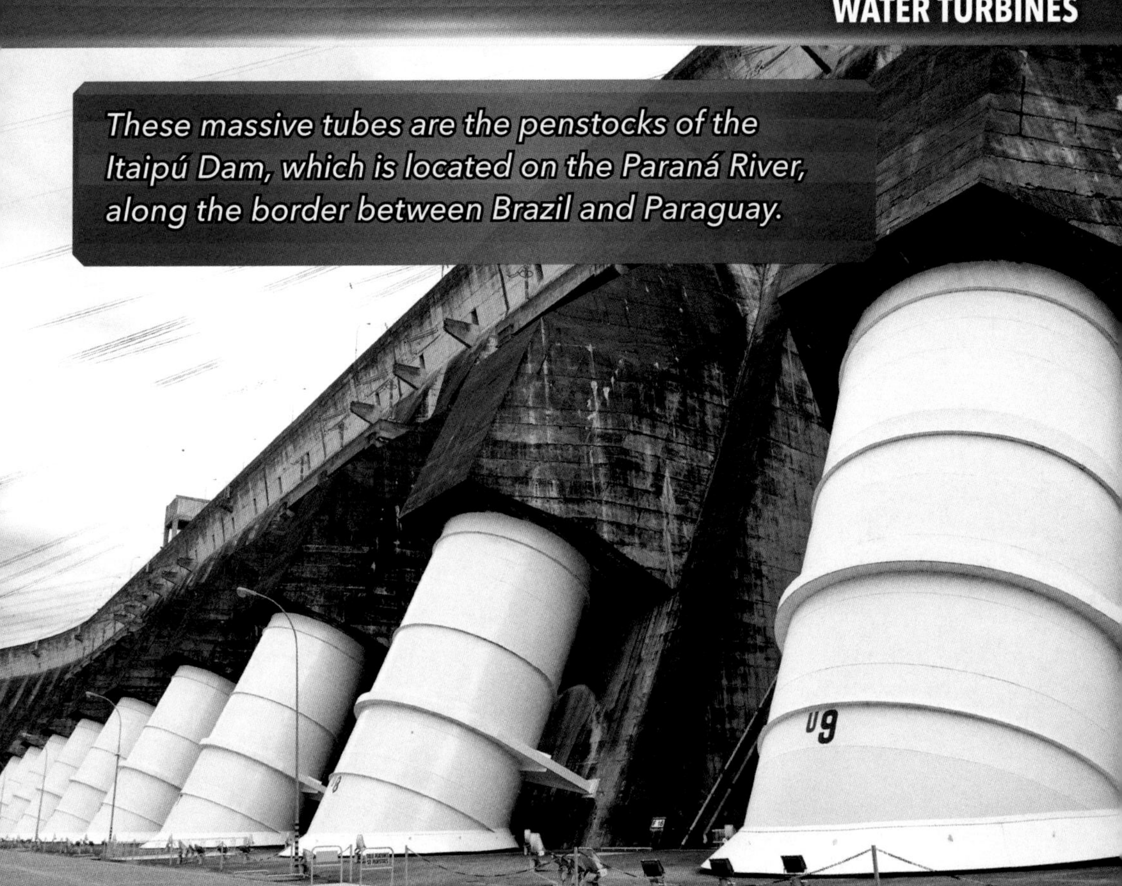

These massive tubes are the penstocks of the Itaipú Dam, which is located on the Paraná River, along the border between Brazil and Paraguay.

turbine. In some applications, both relief valves and surge chambers have been used.

HOW WATERPOWER IS MEASURED

The power produced by water depends upon the water's weight and head (height of fall). Each cubic foot of water (0.03 cubic meter) weighs 62.4 pounds (28.3 kilograms). For example, a column of water that is 1 foot square (0.3 meter square) and 10 feet (3 m)

TURBINE MODEL TESTING

Before building large-scale installations, the design should be checked out with turbine model tests, using geometrically similar models of small and intermediate size, all operating at the same specific speed. Allowances must be made for the effects of friction, determined by the Reynolds number (density × rotational speed × runner diameter squared/ viscosity) and for possible changes in scaled roughness and clearance dimensions. Friction effects are less important for large units, which tend to be more efficient than smaller ones.

high would contain 10 cubic feet (0.28 cu m) of water. It would press upon each square foot of turbine blade with a force of 624 (10 × 62.4) pounds (283 kg).

Engineers measure waterpower in terms of horsepower. One horsepower is the force it takes to raise 33,000 pounds (14,970 kg) one foot in one minute, or 550 pounds (250 kg) one foot in one second. The horsepower potential of a waterfall is found by multiplying its flow, measured in cubic feet per second, by its height, measured in feet. Then the product is multiplied by

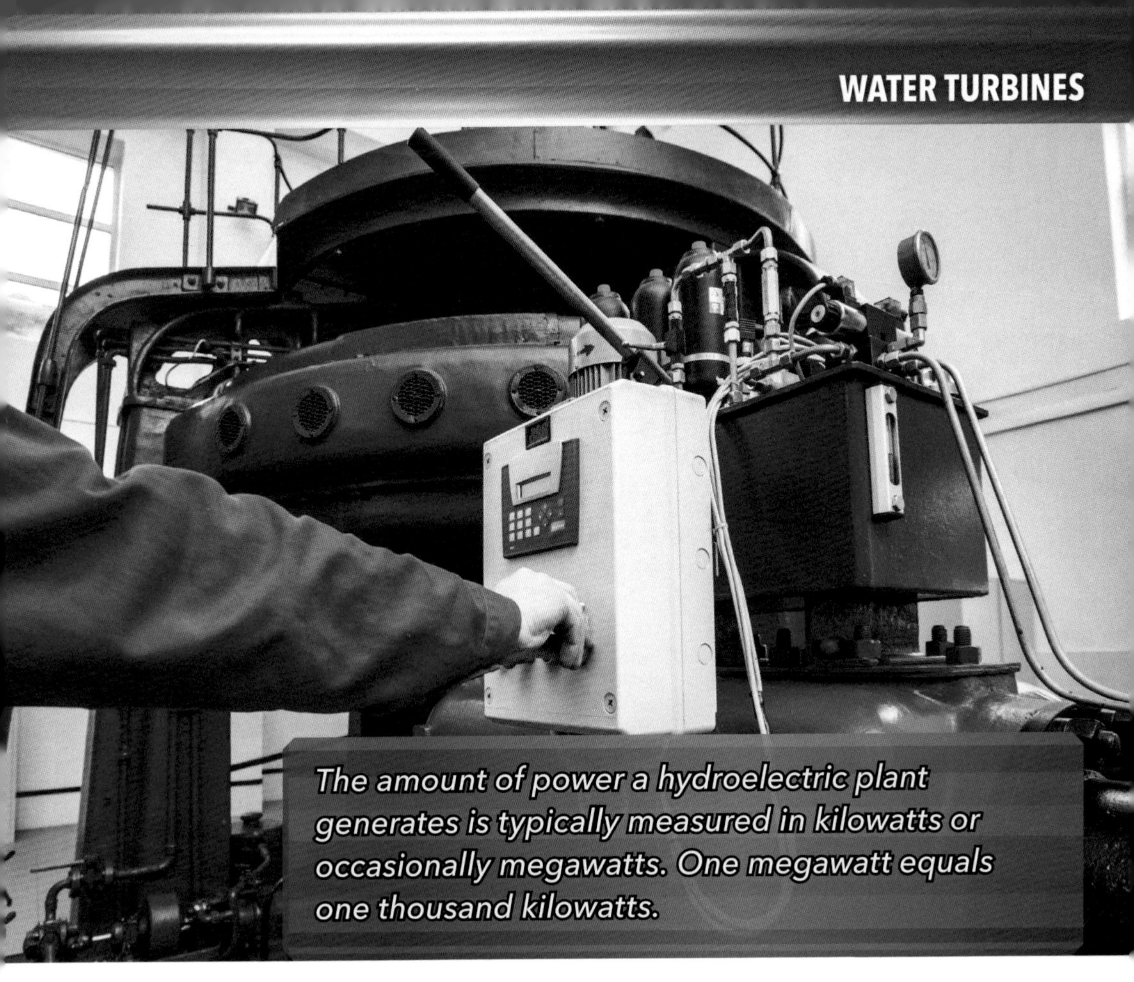

The amount of power a hydroelectric plant generates is typically measured in kilowatts or occasionally megawatts. One megawatt equals one thousand kilowatts.

0.113, which is 62.4 (the number of pounds in a cubic foot of water) divided by 550. A 10-foot- (3 m) high waterfall with a flow of 100 cubic feet (2.83 cu m) per second would develop 100 × 10 × 0.113, or 113, horsepower.

The output of a hydroelectric plant is usually measured in kilowatts of electricity. One horsepower equals 0.746 kilowatt. Thus, a hydroelectric turbine that develops 65,000 horsepower has a capacity of 65,000 × 0.746, or 48,490, kilowatts of electricity.

CHAPTER THREE

HYDROELECTRICITY HISTORY

By the mid-nineteenth century, water turbines were widely used to drive sawmills and textile mill equipment, often through a complex system of gears, shafts, and pulleys. After the widespread adoption of the steam engine, they ceased to be a major factor in power generation until the advent of the electric generator made hydroelectric power possible.

TURBINE DEVELOPMENTS

In 1826 Jean-Victor Poncelet of France proposed the idea of an inward-flowing radial turbine, the direct precursor of the modern water turbine. This machine had a vertical spindle and a runner with curved blades that was fully enclosed. Water entered radially inward and discharged downward below the spindle.

A similar machine was patented in 1838 by Samuel B. Howd of the United States and built subsequently. Howd's design was improved on by James B. Francis, who added stationary guide vanes and shaped the blades so that water could enter shock-free at the correct angle. His runner design, which came to be known as the Francis turbine, is still the most widely used for medium-high heads. Improved control was proposed by James Thomson, a Scottish engineer, who added coupled and pivoted curved guide vanes to assure proper flow directions even at part load.

Jean-Victor Poncelet distinguished himself both as a mathematician and as an engineer. He applied mathematics to the improvement of turbines and waterwheels.

Francis turbines were augmented by the development of the Pelton wheel (1889) for small flow rates and high heads and by propeller turbines, first built by Victor Kaplan in 1913, for large flows at low heads. Kaplan's variable-pitch propeller turbine, which still bears his name, was manufactured after 1920. These units, together with the Deriaz mixed-flow turbine (invented by Paul Deriaz in 1956), constitute the arsenal of modern water turbines.

HYDROELECTRICITY PLANTS

The world's first hydroelectric central station was built in 1882 in Appleton, Wisconsin, only three years after Thomas Edison's invention of the light bulb. Its output of 12.5 kilowatts was used to light two

HOOVER DAM

Constructed between 1930 and 1936, Hoover Dam is the highest concrete arch dam in the United States. It impounds Lake Mead, which extends for 115 miles (185 km) upstream and is one of the largest artificial lakes in the world. The dam is used for flood and silt control, hydroelectric power, agricultural irrigation, and domestic water supply.

Hoover Dam is 726 feet (221 m) high and 1,244 feet (379 m) long at the crest. Four reinforced-concrete intake towers located above the dam divert water from the reservoir into penstocks. The water, after falling some 500 feet (150 m) through the pipes to a hydroelectric power plant at the base of the dam, turns 17 Francis-type vertical hydraulic turbines, which rotate a series of electric generators that have a total power capacity of 2,080 megawatts. Nearly half of the generated electric power goes to the Metropolitan Water District of Southern California, the city of Los Angeles, and other destinations in southern California; the rest goes to Nevada and Arizona.

The Hoover Dam spans Black Canyon, harnessing the Colorado River and impounding Lake Mead.

paper mills and a house. Thereafter hydroelectric power development spread rapidly, though even by 1910 most units delivered only a few hundred to a few thousand kilowatts. Installations with more than 100,000-kilowatt capacity were not built until the 1930s. One of the first large US plants was installed at Hoover Dam on the Colorado River between Nevada and Arizona. It began operating in 1936 and eventually included 17 Francis turbines capable of delivering from 40,000 to 130,000 kilowatts of power, along with two 3,000-kilowatt Pelton wheels.

The first pumped storage plant with a capacity of 1,500 kilowatts was built near Schaffhausen, Switzerland, in 1909. It made use of a separate pump and turbine, resulting in a relatively large and only barely economical system. The first US plant, built on the Rocky River in Connecticut in 1929, was also only marginally economical. In the United States major work on pumped-storage hydropower began in the mid-1950s, following the success of a plant at Flatiron, Colorado. Built in 1954, this facility was equipped with a reversible-pump turbine having a capacity of 9,000 kilowatts.

In highly industrialized countries, such as the United States and the nations of western Europe, most potential sites for hydropower have already

To even the load on the generators, pumped-storage hydroelectric stations are occasionally built. Pumped-storage systems are efficient and provide an economical way to meet peak loads.

been tapped. Environmental concerns relating to the impact of large dams on the upstream watercourse and to the possible effect on aquatic life add to the likelihood that only a few large hydraulic plants will be built in the future.

From about the 1940s to the early 1970s, many small US hydroelectric facilities (primarily those of less than 1,000-kilowatt capacity) were, in fact, closed down because high maintenance and supervision costs made them uneconomical compared to power plants that burn fossil fuels. Even though the increase in fossil-fuel costs since 1973 has led to the rehabilitation of some of these abandoned plants, only a marked increase in fuel prices, coupled with specific needs for irrigation or flood control, is likely to lead to significant new hydroelectric plant construction.

Since 1930 most of the dams in the United States have been erected through federal or local

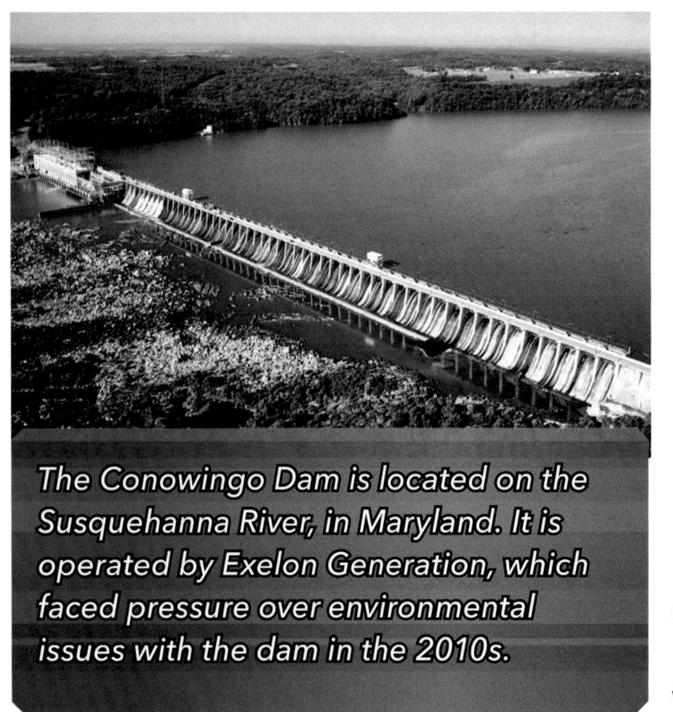

The Conowingo Dam is located on the Susquehanna River, in Maryland. It is operated by Exelon Generation, which faced pressure over environmental issues with the dam in the 2010s.

agencies. However, private utility companies also have major plants on the Columbia River, on the Susquehanna in Maryland and Pennsylvania, on the Connecticut River in New England, and on the Saluda River in South Carolina. Waterpower development by federal agencies is the responsibility of the United States Army Corps of Engineers, the Bureau of Reclamation, or the Tennessee Valley Authority (TVA). Except for TVA projects, federal waterpower developments are funded by selling electricity produced at federal dams that were originally built primarily for flood control.

As of 2016, the U.S. Energy Information Administration (EIA) estimates that hydropower provides about 6.5 percent of the utility-scale electricity generation in the United States. That makes it the renewable energy source that produces the most electricity.

CHAPTER FOUR

UPSIDES AND DOWNSIDES

As all power sources do, hydroelectricity has both advantages and disadvantages. Many of the negative environmental impacts of hydroelectric power come from the associated dams, which can interrupt the migrations of spawning fish, such as salmon, and permanently submerge or displace ecological and human communities as the reservoirs fill.

CLEAN, RENEWABLE ENERGY

Falling water is one of the three principal sources of energy used to generate electric power, the other two being fossil fuels and nuclear fuels. Hydroelectric power has certain advantages over these other sources: it is continually renewable owing to the recurring nature of the water cycle and produces neither atmospheric nor thermal pollution. Fossil fuels, on the other hand, produce gases that contribute to both pollution and global warming. There is also a limited—and quickly diminishing—supply of them. Objections to nuclear power include safety concerns and worries about how and where to safely store the radioactive waste that it produces.

In the early twenty-first century, waterpower accounted for roughly a fifth of the world's total electricity production. Along with China and Brazil, the United States, Canada, and Russia were among the top hydroelectric-generating countries. In many countries around the world, waterpower was the only renewable energy source in use, and in several of them—including Lesotho, Paraguay, and Zambia—hydroelectric plants supplied virtually all of the electrical power.

LOCATION LIMITATIONS

While hydroelectricity provides a supply of clean energy that cannot be used up, it does not make sense in every location. Hydroelectric power is a preferred energy source in areas with heavy rainfall and with hilly or mountainous regions that are in reasonably close proximity to the main load centers. Some large hydro sites that are remote from load centers may be sufficiently attractive to justify the long high-voltage transmission lines. Small local hydro sites may also be economical, particularly if they combine storage of water during light loads with electricity production during peaks.

Itaipú Dam, on the Paraná River bordering Brazil and Paraguay, has eighteen huge turbines that together can produce more than 12,500 megawatts of electricity.

Although large hydroelectric plants can be operated economically, the cost of land acquisition and of dam and reservoir construction must be included in the total cost of power, since these outlays generally account for about half of the total initial cost. Most large plants serve multiple purposes: hydropower generation, flood control, storage of drinking water, and the impounding of water for irrigation. If the construction costs are properly prorated to the non-power-producing utility of the unit, electricity can be sold very cheaply. In the Pacific Northwest

The Bonneville Dam, on the Columbia River, was completed in 1937. It was a project of the Public Works Administration, which built multiple dams during the Great Depression of the 1930s.

region of the United States, such accounting has given hydroelectric plants an apparent cost advantage over fossil-fueled units.

PROBLEMS WITH DAMS

If a dam fails, or breaks, it can cause devastating floods downstream that may kill many people. But working dams also may have a variety of negative effects on the environment or on the people and

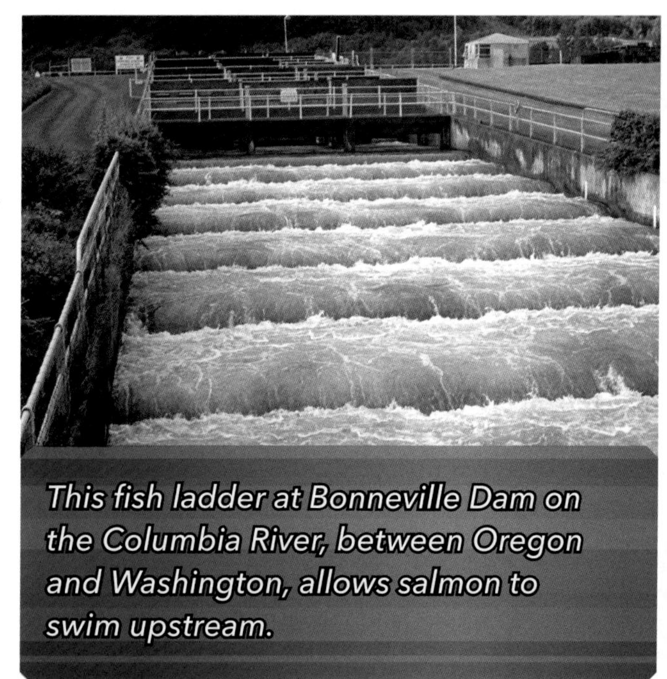
This fish ladder at Bonneville Dam on the Columbia River, between Oregon and Washington, allows salmon to swim upstream.

animals in the region they are located.

Dams may impair fishing, though structures can be built to help solve the problem. For example, at Bonneville Dam (1937) on the Columbia River in Oregon, fishways help the salmon around the dam as they swim upstream to spawn. Fish ladders enable them to jump from one ascending pool to another. Fish locks lift the fish like locks lift ships.

Yet another problem of dams is silting. Some rivers pick up clay and sand and deposit them behind the dam and thereby lessen its usefulness. This process may also rob the soil of its natural fertility and reduce populations of fish that depend on the silt's nutrients.

An additional objection is the fact that reservoirs may cover towns or historic and scenic places. In the 1960s engineers and archeologists relocated

GLEN CANYON DAM

Built between 1956 and 1966, the Glen Canyon Dam eventually came to be seen by environmentalists as being responsible for destroying a vast, beautiful pristine landscape. Anger over the Glen Canyon Dam energized the Sierra Club to mount a major campaign against additional dams proposed for construction along the Colorado River near the borders of Grand Canyon National Park. By the late 1960s, plans for these proposed Grand Canyon dams were politically dead. Although the reasons for their demise were largely the result of regional water conflicts between states in the Pacific Northwest and states in the American Southwest, the environmental movement took credit for saving America from the desecration of a national treasure.

an ancient Egyptian temple complex that would have been submerged by rising waters caused by the Aswan High Dam. Between 1963 and 1968 a workforce and an international team of engineers and scientists dug away the top of the cliff and took apart both temples, sawing them into giant blocks. These were hoisted to the top of a cliff, in the face of which they were originally carved, for reassembly.

In China the Three Gorges Dam (constructed from 1994 to 2006) generated significant opposition within China and in the international community. Millions of people were displaced by, and cultural and natural treasures were lost beneath, the reservoir that was created following erection of the 607-foot- (185 m) high concrete wall, some 7,500 feet (2,300 m) long, across the Yangtze River. The dam is capable of producing 22,500 megawatts of electricity (which can reduce coal usage by millions of tons per year), making it one of the largest hydroelectric producers in the world.

The temples at Abu Simbel were rebuilt on high ground more than 200 feet (60 meters) above the original location. In all, some 16,000 blocks were moved.

GLOSSARY

AXIAL Situated around, in the direction of, on, or along an axis.

ESTUARY Where the tide meets a river current.

GENERATOR A machine by which mechanical energy is changed into electrical energy.

GLOBAL WARMING A warming of Earth's atmosphere and oceans that is widely thought to result from an increase in the greenhouse effect caused by air pollution.

HYDRAULIC Operated, moved, or brought about by means of water.

INERTIA A property of matter by which it remains at rest or in unchanging motion unless acted on by some external force.

INLET An opening for intake, especially of fluids.

IRRIGATION The artificial supply of water to agricultural land.

KINETIC ENERGY Energy associated with motion.

LATITUDE The distance north or south from the equator, as measured in degrees.

MECHANICAL ENERGY All the energy that an object has because of its motion and its position. Mechanical energy is equal to kinetic energy plus potential energy.

OSCILLATE To move or travel back and forth between two points.

POTENTIAL ENERGY The amount of energy a thing (as a weight raised to a height or a coiled spring) has because of its position or because of the arrangement of its parts.

RADIAL Arranged or having parts arranged like rays around a common center.

RESERVOIR An artificial or natural lake where water is collected as a water supply.

ROTOR A part that rotates in a stationary part (as in an electrical machine).

RUNNER A thin piece or part on which something slides.

TURBINE An engine whose central driving shaft is fitted with a series of blades spun around by the pressure of a fluid (as water, steam, or air).

SPAWN To deposit or fertilize eggs.

SPILLWAY A passage for extra water to run over or around a dam.

SPINDLE Something (as an axle or shaft) shaped or turned like a spindle or on which something turns.

THERMAL POLLUTION The release of heated liquid (as water used by a factory) into a natural body of water at a temperature harmful to the environment.

VISCOSITY The resistance of a fluid (liquid or gas) to a change in shape or movement of neighboring portions relative to one another.

FOR MORE INFORMATION

Bailey, Diane. *Hydropower* (Harnessing Energy). Mankato, MN: Creative Education, 2015.

Bard, Jonathan. *Hydroelectricity: Harnessing the Power of Water* (Powered Up! A STEM Approach to Energy Sources). New York, NY: PowerKids Press, 2018.

Bethea, Nikole Brooks. *Building Dams* (Engineering Challenges). Lake Elmo, MN: Focus Readers, 2018.

Bjorklund, Ruth. *The Pros and Cons of Hydropower* (Economics of Energy) New York, NY: Cavendish Square, 2015.

Brundle, Harriet. *Renewable Energy*. New York, NY: KidHaven Publishing, 2018.

Dickmann, Nancy. *Harnessing Hydroelectric Energy* (The Future of Energy). New York, NY: PowerKids Press, 2017.

Dickmann, Nancy. *Harnessing Wave and Tidal Energy* (The Future of Energy). New York, NY: PowerKids Press, 2017.

Dickmann, Nancy. *Water: Hydroelectric, Tidal, and Wave Power* (Next Generation Energy). New York, NY: Crabtree Publishing Company, 2016.

Doeden, Matt. *Finding Out about Hydropower* (Searchlight Books: What Are Energy Sources?).

Minneapolis, MN: Lerner Publications Company, 2015.
Goldish, Meish. *The Hoover Dam* (American Places: from Vision to Reality). New York, NY: Bearport Publishing, 2017.
Grady, Colin. *Hydropower* (Saving the Planet through Green Energy). New York, NY: Enslow Publishing, 2017.
Jennings, Terry Catasús. *Ocean Energy* (Core Library: Alternative Energy) Minneapolis, MN: ABDO Publishing, 2017.
Murray, Laura. *Hydroelectric Energy* (Core Library: Alternative Energy) Minneapolis, MN: ABDO Publishing, 2017.
Nagelhout, Ryan. *Dams* (Technology Takes on Nature). New York, NY: Gareth Stevens Publishing, 2017.
Simon, Seymour. *Water: All About the Water Cycle, Precipitation, Why We Need Water, and More!* New York, NY: Harper, 2017.
Sneideman, Joshua, and Erin Twamley. *Renewable Energy: Discover the Fuel of the Future with 20 Projects* (Build It Yourself). White River Junction, VT: Nomad Press, 2016.
Spilsbury, Louise. *Dams and Hydropower* (Development or Destruction?) New York, NY: Rosen Central, 2012.

WEBSITES

Bureau of Reclamation
https://www.usbr.gov
Facebook: @bureau.of.reclamation, Twitter: @usbr

U.S. Department of Energy
https://energy.gov/eere/water/water-power
-technologies-office
Facebook: @energygov; Instagram, Twitter: @energy

The USGS Water Science School
https://water.usgs.gov/edu/hyhowworks.html
Facebook: @USGeologicalSurvey;
Instagram, Twitter: @USGS

World Energy Council
https://www.worldenergy.org/data/resources/resource/
hydropower
Twitter: @WECouncil

INDEX